NASA MARS MISSION

FOR KIDS

A SPACE BOOK OF FACTS, ACTIVITIES, AND FUN

Woo! Jr. Kids Activities & Wendybird Press Founder: Wendy Piersall
Text Written & Researched by Kate Whitcomb
Art Director: Lilia Garvin
Production Designer: Ethan Piersall
Cover Illustration: Michael Koch | Sleeping Troll Studios www.sleepingtroll.com
Editing & Proofreading: Lori Acosta

Published by:
Wendybird Press
226 W. Judd
Woodstock IL, 60098
www.wendybirdpress.com

ISBN-13: 978-1732958982
ISBN-10: 1-73295898X

Find more Screen-Free puzzles in:
Big Brain Teasers Book for Kids

This huge book of 360+ brain teaser puzzles for kids is perfect for ages 9 - 12 and up. Included are long time family favorite mind teasers such as hidden pictures, cryptograms, math squares, logic grid puzzles, picross and matchsticks. Also included are cool Japanese puzzles like sudoku, masyu, slitherlink, and numberlink.

Brain teasers can:

- Boost brain power
- Improve concentration
- Develop short term memory competency
- Cultivate problem solving skills
- Promote critical thinking abilities

Enjoy this children's puzzle book on school breaks, while you travel, or any day you need some screen-free fun mental exercise!

Purchase on Amazon at:
amzn.to/38Gz67G

Destination: MARS

Introducing: MARS

Humans have been able to see Mars, also known as the Red Planet, with the naked eye since ancient times. In 1997, NASA sent the first Mars Rover, Sojourner, to travel the surface of Mars on the Pathfinder mission. Sojourner landed in the Jezero crater and was designed to take photographs and find evidence of life on Mars. With the new information gained by the first Mars Rover, scientists were able to send Spirit and Opportunity to find evidence of current or past sources of water. It is important for scientists to observe the environment of the Martian surface and how it could affect life forms like humans. Radiation from the Sun, for instance, is dangerous to humans and the difference in the force of gravity, or the g-force, could have negative effects on human health. The continuing exploration of Mars will hopefully help scientists find solutions to these obstacles.

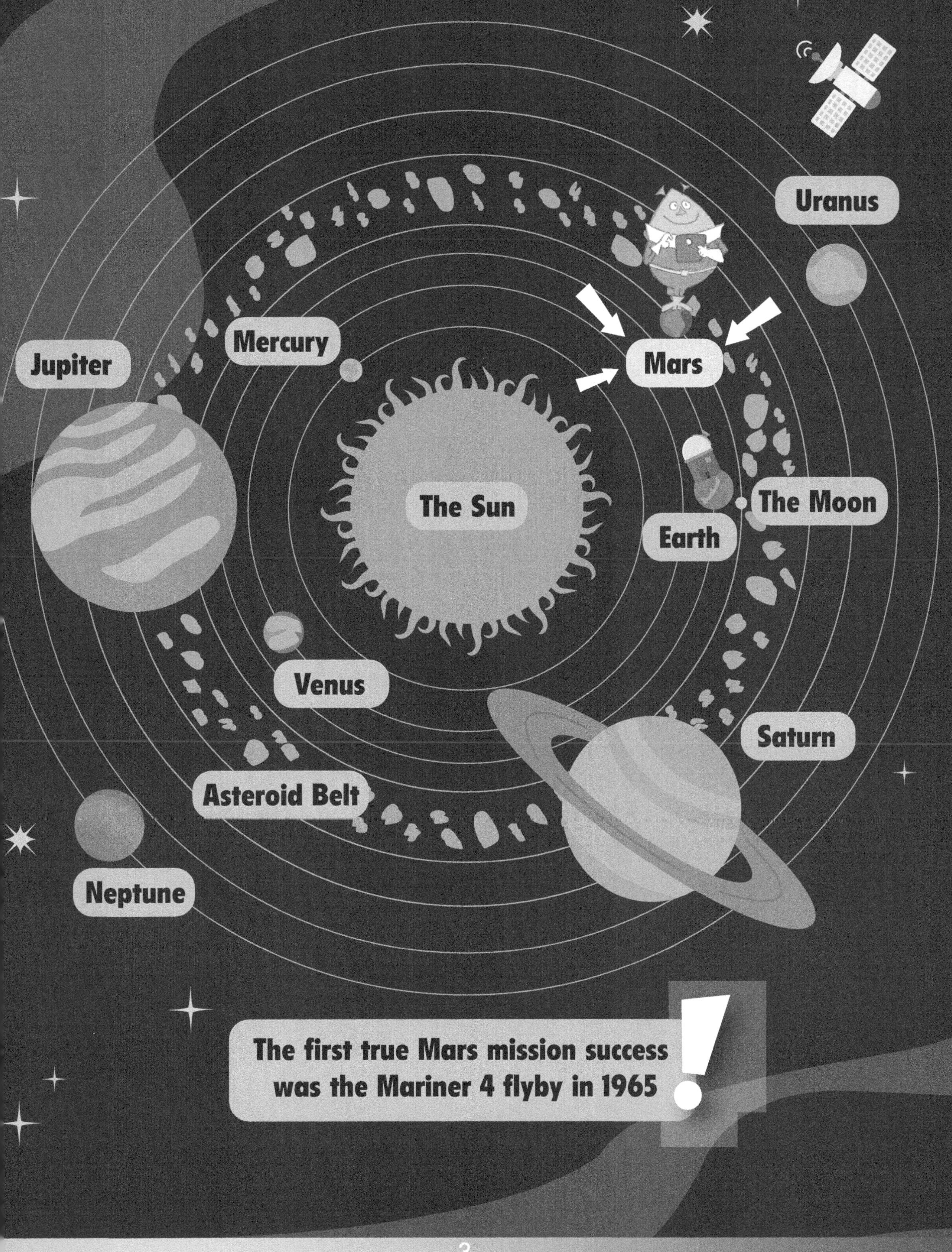

Uranus
Mercury
Jupiter
Mars
The Sun
The Moon
Earth
Venus
Saturn
Asteroid Belt
Neptune
The first true Mars mission success was the Mariner 4 flyby in 1965

Mission Overview: Mars 2020

Scientists think it's possible that there has been life on Mars before, but they still have a lot of questions about it. Mars 2020 is the official name of the NASA mission to discover evidence of past life on the planet Mars, which will use the Perseverance rover. Perseverance will land in a specific place on Mars called the Jezero crater and travel the planet's surface looking for answers to many scientific questions. There have been four rovers sent to Mars before, but this one will have new goals and scientific instruments. Mars rovers are more helpful to scientists than stationary landers, because they can examine more of the planet's surface, position themselves in sunny places to store solar power, and be directed by astronauts on Earth.

Fun Fact!

The Jezero Crater, a crater on the surface of Mars, is thought to once have held water, and its name comes from the Slavic word for "lake."

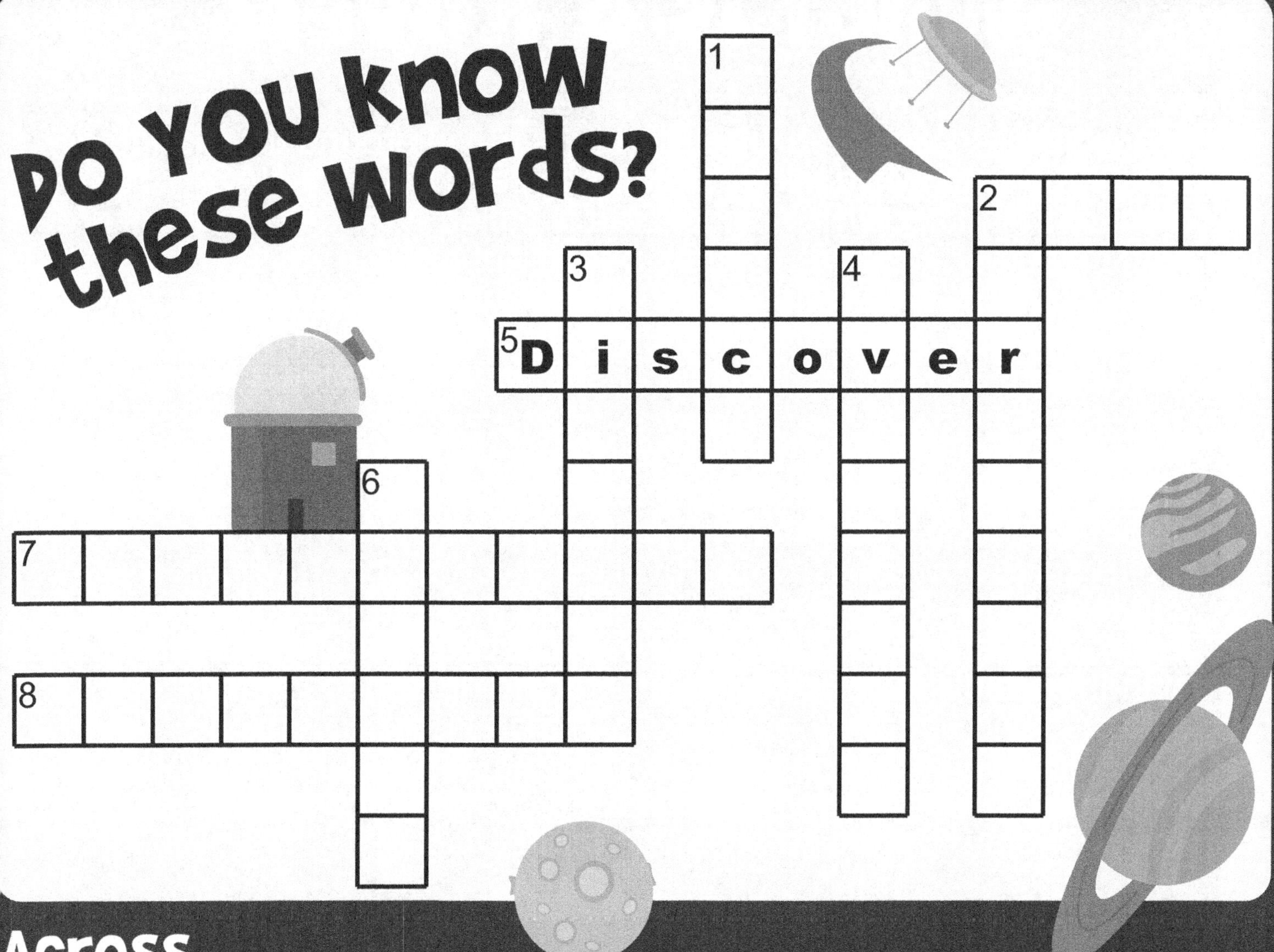

Across

2. The fourth planet from the Sun, named after the ancient Roman god of war and agriculture
5. To gain new knowledge
7. The investigation of unknown regions
8. Energy from the Sun that gives us visible light and sometimes sunburns

Down

1. The force of gravity or acceleration on a body
2. A motor vehicle designed to land on and travel around the surface of Mars
3. An important task that has been assigned
4. Something that proves or disproves a theory; proof
6. A bowl-shaped depression on the planet's surface caused by a volcano or the impact of something large like a meteor

What is a Rover?

NASA has sent several kinds of spacecraft to Mars. One kind of spacecraft is called an **orbiter**. Orbiters travel around the planet, or orbit the planet, while taking pictures to send back for NASA's research. Another kind of spacecraft is called a **lander**. Landers are sent to the surface of the planet, but do not move from their landing spots. Landers take pictures like orbiters but also observe and measure other information, like the temperature or wind speed on the planet's surface. A third kind of spacecraft is called a **rover**.

Including the Perseverance rover, NASA will have sent 5 rovers to Mars. Rovers travel over the planet's surface, or rove around the planet's surface, and send information back to NASA just like orbiters and landers. Rovers travel using large, tough wheels that can handle the rocky surface better than most vehicles we see on Earth, like cars and trucks. They travel to different areas of the planet, not only taking pictures and recording surface details, but also studying and testing the rocks, minerals, and chemicals they come across. This information is important to scientists at NASA, who hope to learn more about past life on Mars and how we could send astronauts there safely in the future.

Can you find the hidden pictures?

Perseverance: Scientific Goals

The Mars Exploration Program uses science to understand whether Mars is, was, or ever could be habitable, which means 'to be capable of sustaining life.' The four general goals of the program will help NASA scientists understand the geology, climate, and other processes that have created Mars and its environment over time and how they might impact humans in the near future.

Goal #1

Find evidence that the environment on Mars was ever capable of sustaining life. The main objective of this mission is to discover evidence that liquid water not only existed on Mars, but that it was also stable and long-lasting enough for life to develop. A possible source of such water might be found near hot spots scientists have already found evidence of. These may be signs of current or inactive hydrothermal pools like those found in Yellowstone National Park. In addition to lasting water, scientists also need to find evidence of other building blocks of life, called biosignatures. For instance, the element carbon is necessary for life on Earth so its presence on Mars could tell scientists about past or current life on the planet. It's likely that the organisms scientists may find evidence of here would be microbial, like bacteria on Earth.

Goal #2

Observe the current climate of Mars to learn about its past. Just like Earth, Mars has seasonal changes and polar ice caps, but it currently has dust storms so large that they can encompass the entire planet. Discovering the causes of these storms and studying the current weather patterns may help scientists understand the history of climate change on Mars. With the combination of detailed weather maps and information about the composition of the atmosphere, scientists hope to understand more about the seasonal changes and thus determine a pattern that explains how Mars has changed over time.

Goal #3

Observe the current geology of Mars to learn about its past. Mars has other similarities with Earth, like having a magnetic field and volcanoes. Studying these features and looking at the composition of the planet's rocks can tell scientists a lot about the roles of wind, water, volcanism, tectonics, cratering and other processes that have shaped the surface of Mars. For instance, the way tectonics, or the movement of a planet's crust, impact the Martian surface is very different than they do on Earth. This difference causes lava to pool into one concentrated area and make massive volcanoes that are between 10 and 100 times larger than any volcano on Earth. Another way to learn about Martian geology is through studying the planet's magnetic fields, which can teach us about the temperature and interior of Mars in the past. Magnetic fields are often meant to shield planets from radiation so their existence implies that Mars may have once been much more like Earth than it is today.

Goal #4

Prepare for humans to explore Mars in the future. Exploring Mars' environment will also help the Perseverance rover send information back to NASA about water and ways to extract oxygen from the atmosphere for astronauts to survive on Mars in future missions. Simply getting astronauts to the Martian surface and back to Earth safely would be extremely difficult, and a thorough understanding of the environment is necessary for both human health and the operation of any necessary technologies. The thin atmosphere of Mars allows a lot of radiation to reach its surface from the Sun, so protective gear and habitats will need to be developed.

Perseverance: In Summary

The goal of the next NASA mission to Mars is to discover whether the planet has ever been habitable for life forms, or if it might even still be safe for life forms to live there now. These life forms, or organisms, were likely to be very small, probably even microbial like Earth's bacteria. By studying weather patterns, scientists will be able to determine the planet's current climate. This will help scientists understand the seasonal changes, which will give us some insight into the history of Mars and how the planet has changed over time. Another source of historical information about Mars is its geology, which scientists have learned about through studying rocks and how its crust was formed and changed through tectonics. Temperature is also an important aspect to study, whether it's through hot spots that may have had hydrothermal pools or by studying magnetic fields to determine the planet's past temperatures. All of this information combined can teach us how to help astronauts safely survive Mars's dangerous atmosphere.

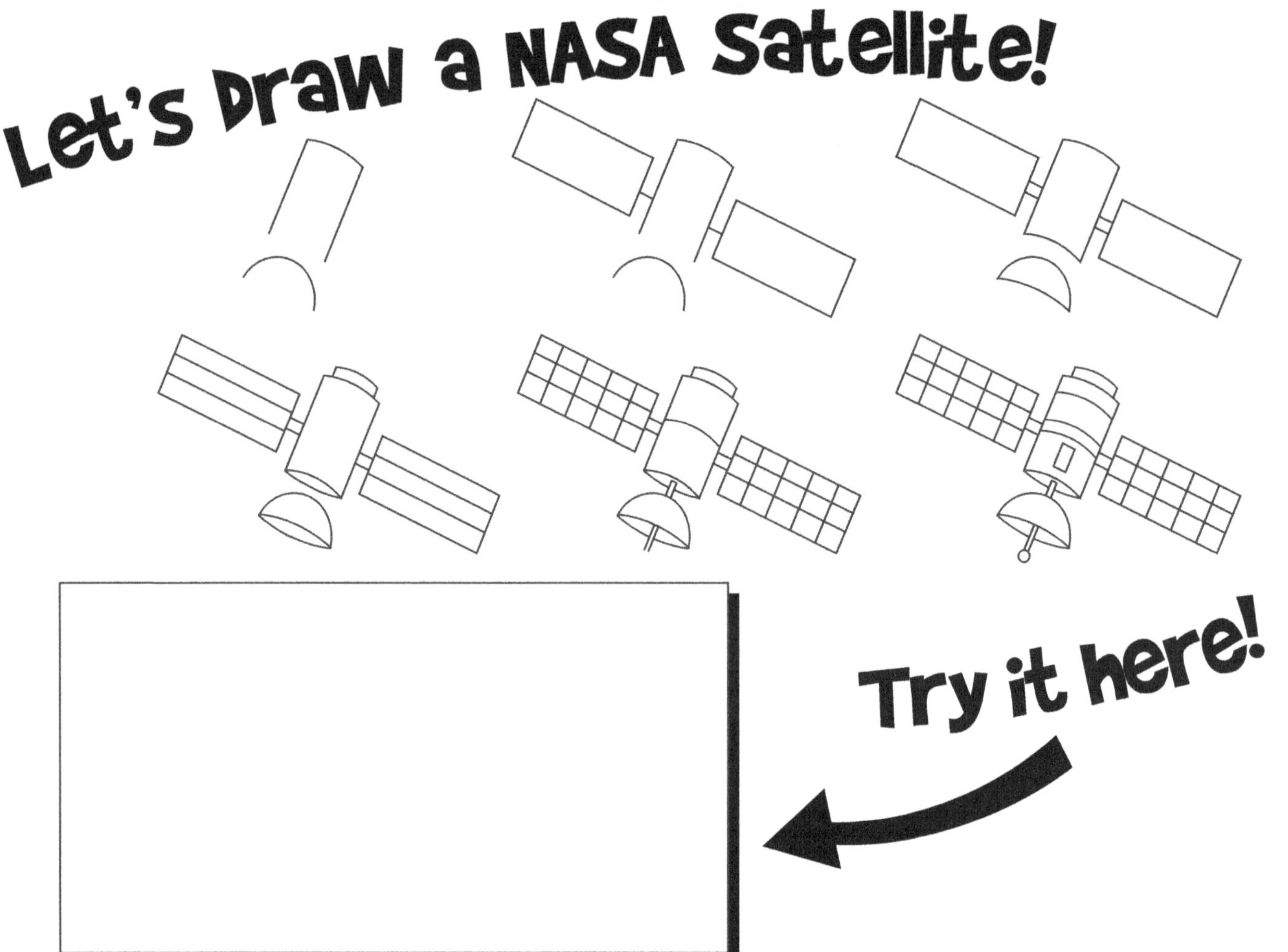

Find your way to the Jezero crater! 
Start HERE

Can you find 5 differences?

Mars Riddles

What is red from far away, but more of a butterscotch up close...and no, it's not a gaseous place, it's full-on terrestrial!?

Answer: Mars.

They call me Iron Oxide, but I prefer my nickname, "Iron Dude". I'm responsible for making Mars what?

Answer: Red.

Hi, I'm Zzzyccskk, and I'm from Mars. I've got cool alien clothes and toys. The years here are much longer than yours on Earth. Since I live here, what am I called?

Answer: A Martian.

Earth only has one of me. Mercury and Venus have none. Jupiter has sixty-three, oh my! And Mars has two. What am I?

Answer: Moons.

We hear your planet has lots of our kind, who spew hot magma, creating islands or scary events for your humans...but listen here, we are much taller than all your "fire mountain" friends...we would kick their butts in an instant! Who are we?

Answer: Volcanoes.

Previous Mars Missions...

The rovers sent to Mars in the past have discovered water and other building blocks for life. Curiosity is the only rover still active, and the one most similar to Perseverance, the newest Mars rover. It was sent to discover chemicals needed to support life on Mars, and whether there was a lasting source of water at some point. Curiosity's mission was based on the earlier mission of the rovers Spirit and Opportunity: to find evidence of water on Mars. The first rover to land on Mars was Sojourner, during the Pathfinder Mission, over 20 years ago. Sojourner was a small, lightweight rover that only had 2 scientific instruments because its mission was simply to explore Mars, take pictures, and discover any important information for building and launching future rovers.

Fun Fact!

Olympus Mons, a mountain on Mars, is the largest mountain in our solar system! It's named after Mount Olympus, the home of the gods in Roman mythology.

Space Exploration Word Search

```
Y B P P F L F T S D R Z N N C F V
R T X F E R Q E U I O M O K J E U
H N I K G R O W F S V N I O U U R
Q I G V G N S H R C E Q T O N S N
W R J L A B A E I O R W A X J A W
E Z Y C Z R F V V V S E I Y B P Z
U K L X E V G Q M E A A D G F M T
K O I D N X A X B R R X A H A N G
V N F U J J C A T M A A R G S S U
N O I S S I M H D F S Q N N U M H
S U U K M K L E J D Y E I C E H Z
Y S K T K R N T T U T X I K E Z W
S L T G E W D A C I F G L N U E J
Q D W U I S D M C I A S V F E H O
S P N V A L H I H M X I N M O P G
A T E B G N D L F F R F J P F T R
P I Y V J L O C L O J C R A T E R
U L D W I G E R N J G A A Y H X U
V M A Y F D P M T J D N K A B P J
I D F N E R E H P S O M T A Y L M
E D Y N E N Z N X F A F C G I O S
S R A M T T R V C Y Y B O R P R R
C H X I L L G G F E V L W J P A K
N R D Q H B D J D T O U Q F D T Q
C T X O Y Q H O Q E T T E F G I S
W R K S X W F P G J C T J I T O L
R L K U E S R E R K E C V M S N M
H A B I T A B L E W Y S K T U X V
```

Astronauts

Atmosphere

Climate

Crater

Discover

Earth

Environment

Evidence

Exploration

Geology

Gravity

Habitable

Magnetic

Mars

Mission

Perseverance

Planet

Radiation

Rover

Volcanoes

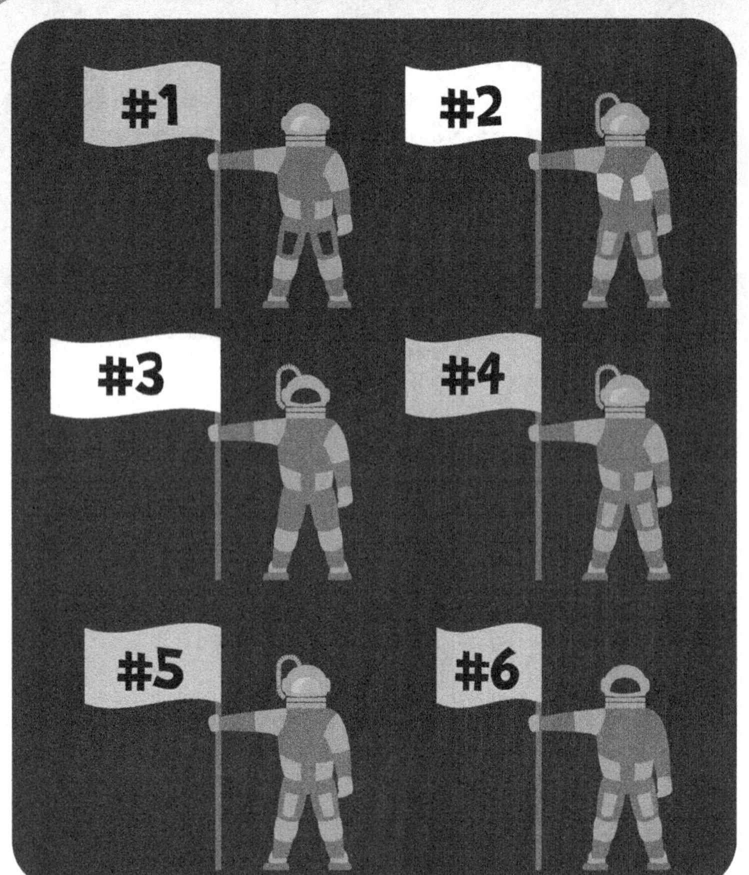

Can you find the pair of astronauts that match?

#___ & #___

+ Future Mars Missions!

Looking forward, NASA hopes to one day send humans to Mars. Through the information gained by the Mars rovers, scientists have learned much about the Red Planet's climate and environment. With the next Mars mission, they hope to also test technologies that could assist astronauts who travel there in the future. Future missions, with both rovers and orbiting structures, will continue to gather information using special scientific tools so scientists will be able to develop new necessary technologies. Some of the dangers to astronauts traveling to Mars include risk of radiation, lack of access to fresh water, and the effects of G-force, defined as the effects of gravity or acceleration. Future orbiting, land-based, and rover missions will focus on finding solutions to questions about how to ensure the survival of astronauts traveling to Mars. The thin atmosphere of Mars allows a lot of radiation to reach its surface from the Sun, so protective gear and habitats will need to be developed. G-force can be incredibly dangerous if it's too great for a human or their spacecraft to survive.

Color in our Solar System!
17

Brad is so excited about Mars...
that he's imagined his own space journey!
Color Brad's imaginary mission!

Discover New Words: A Puzzle

Your mission, should you choose to accept it, is to discover how many new words you can create from the name of the newest NASA Mars rover:

PERSEVERANCE

1 POINT

2 POINTS

3 POINTS

5 POINTS

HOW MANY POINTS DID YOU GET?

\#________

Technology

Each Mars mission, whether it involves a rover, a lander, or an orbiting spacecraft, is meant to continue the development of technology for future missions. Each new idea or device is part of NASA's technological innovation. This allows NASA and other space programs to continue pushing what is possible in space exploration, and even in technologies that can be used here on Earth. This results in a kind of innovation chain: each mission depends on technology developed by past missions in order to develop new technology for future missions.

With the ultimate goal of human exploration, missions to Mars have discovered new information and developed new technologies as a result. This search for information is constantly in motion: the 2001 Mars Odyssey Mission is discovering more about the radiation on Mars, while an orbiting spacecraft is continuing to search for water resources that could support astronauts exploring Mars. If water is discovered, NASA plans to send robots, drills, and more rovers to access the water for future human exploration. The dangers of entering Mars's atmosphere and landing on its surface will be studied in order to develop spacecrafts that reduce the G-forces on astronauts and their spacecrafts. The ongoing research into the technologies below will assist NASA scientists in making the human exploration of Mars possible.

Discover these terms & why they're important

Propulsion: moving something forward
for providing the energy to get to Mars and conduct long-term studies

Power: to supply something with energy
for providing more efficient and increased electricity to the spacecraft and its subsystems

Telecommunications: sending information over a great distance
for sending commands and receiving data faster and in greater amounts

Avionics: electronics used on aerospace vehicles
for operating the spacecraft and its subsystems

Software Engineering: designing and building computer programs
for providing the computing and commands necessary to operate the spacecraft and its subsystems

A Martian Funny Fill-In Story

Fill in the story with the type of words asked for, then read it for a laugh!

A __________ and fast-moving dust storm gathered on the horizon. That was when you,
(adjective)

me, and ____________ decided to head back to the Kingdom of _________. Mars had
(Name) (vegetable)

only been populated with humans for the past thirty years, which is why the giant,

_________ aliens still roamed about. We flew past them all, faster than _______, up over
(noun) (gooey substance)

the Red Sand Cliffs and through the _________ tundra. We thought everything was good,
(something gross)

until we saw it feasting on __________ at the bottom of a massive crater...the terrifying,
(candy)

mysterious, fire-breathing Galactic ____________. That was when I got scared. Not you,
(animal)

though. You said, "Don't worry, _________, I got this! I'm the _______ kid in Marsatonia, and
(Name) (adjective ending in -est)

I will slay this creature with my own _______!" You whirled your sand-cruiser faster than a
(body part)

_________. The next thing I knew, the beast was slayed, and Marsatonia was once again
(food)

safe. You really are a _________, _______ genius. I can't wait to continue our adventures in
(color) (adjective)

the future, here in our awesome new neighborhood beside _______ Beach. Don't forget
(something gross)

about me when you eat _________ donuts and spacedust _________. Okay, that's enough
(type of bug) (type of drink)

out of me... back to cleaning up _________ poop before the Martian Mayor gets mad
(ocean animal)

at me again.

Take Care,
Your Friend,

____
(Funny name)

Red Planet Facts

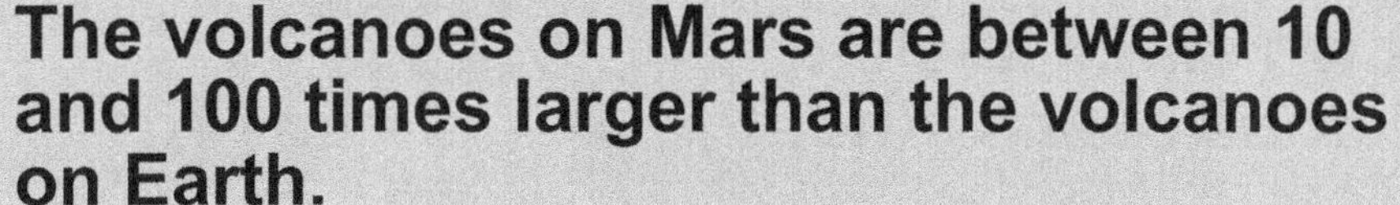

The volcanoes on Mars are between 10 and 100 times larger than the volcanoes on Earth.

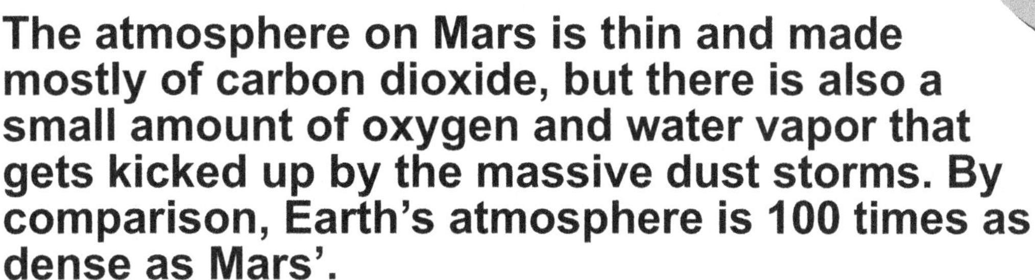

The atmosphere on Mars is thin and made mostly of carbon dioxide, but there is also a small amount of oxygen and water vapor that gets kicked up by the massive dust storms. By comparison, Earth's atmosphere is 100 times as dense as Mars'.

One year on Mars takes almost twice as long as a year on Earth — 687 Martian days instead of 365 Earth days — because it takes so much longer to travel around the Sun. Interestingly, one day on Mars also lasts roughly 24 hours.

The other spacecrafts planning to travel to Mars in 2020: the ExoMars rover (European Space Agency), the United Arab Emirates Hope Orbiter, a Japanese orbiter, a Chinese rover and a SpaceX Dragon capsule. Some may reschedule to 2022, the next launch window.

Mars is known as the Red Planet because iron minerals in the Martian soil oxidize, or rust, causing the soil and atmosphere to look red.

Red Planet Facts

Mars has been known since ancient times because it can be seen without advanced telescopes.

Mars has an active atmosphere, but its surface is no longer active, so its volcanoes are dead. However, one of its volcanoes named Olympus Mons is actually the largest one in our entire solar system.

The Gusev Crater is a crater on the surface of Mars, thought to be where rivers may have led to a massive lake in the past.

The Meridiani Planum is a flat, even plane on the surface of Mars, very close to its equator; "meridiani" means halfway around a circle or sphere, and is related to the name for Earth's Prime Meridian.

12 meteorites from Mars have been found on Earth.

MARS COLORING GRID

Use the color chart included below to color each box!
When you're done you'll have a *complete* picture you colored yourself!

Orange	A(1-3, 7, 12-15), B(1-2, 4, 9-16), C(1-3, 7-16), D(1-8, 11-13, 16-18), E(2-5, 8-10, 12-13, 16-19), F(3-5, 9, 17-18), G(3-6, 9, 11), H(5-8, 10, 12-13), I(8, 11-13), S(18)
Gray	A(4-6, 8-11, 16-20), B(3, 5-8, 17-20), C(4-6, 17-20), D(9-10, 19-20), E(11), F(10-11), G(10, 12), H(9, 11), I(9-10), L(17), M(16)
Blue	M(11-14), N(10-13, 15), O(9-10, 12-13), P(9-13), Q(9-16), R(9, 12, 14, 16), S(10)
Black	E(1), F(1-2), G(1-2), H(1), I(5-7, 14, 16-17, 19), J(5, 7-8, 16-19), K(1, 4-7, 14-20), L(1-2, 4-6, 14-16, 18), M(1-3, 15, 20), N(2, 4, 16, 19-20), O(1, 3-5, 17, 19), P(1, 3, 17-18), Q(4, 6, 17-18), R(3-7), S(6-9), T(4-5, 7, 9-10)
Purple	E(20), F(19-20), G(3, 18-20), H(2-4, 17-20), I(1-4, 15, 18), J(1-4, 9-14, 20), K(3, 8-13), L(3, 7-11, 13, 19-20), M(4-6, 8-10, 19), N(1, 5-9, 18), O(2, 6-8, 18), P(2, 4-8, 19-20), Q(1, 3, 7-8, 19), R(1-2, 8, 17, 19-20), S(1-2, 4-5, 16, 20), T(1-3, 6, 15-17, 19-20)

Would you rather this...

You've awoken in a foreign land, surrounded by auburn dirt swirling into a massive dust storm. To your right, a spaceship awaits. On your left, a family of Martians holds out a sign, *"Please stay with us, we'll teach you awesome things!"* Would you rather hop into the spaceship and head back home in a jiffy, or go hang with the Martians and become a legend?

Would you rather climb the tallest volcano on Mars for *three million dollars* (be careful, don't burn your tush!)...or, ride a *futuristic sand-cruiser mobile* through a wild dust storm for *one* million dollars? Be careful with this choice...according to legend, the Mars volcanoes are littered with stink bugs bigger than your house! And the dust storms? Oh boy...let's just say you better be prepared to get a ton of glowing booger-starved gnats lodged up your nose during the ride!

...or that?

A secret agent just called you. They say you're the only one who can save the galaxy...the only kid bearing a special power within. But in order to save the universe, you must travel to Mars and defeat the all-powerful Gigavonbuttbreath. Would you rather embrace your newly found special powers and head to Mars to become a hero, or tell the secret agent to buzz off (because hey, Giga-butt dude sounds kinda scary)?

The world just announced it's now safe to fly to Mars. Families all over the world are packing their suitcases, ready for the most epic vacation of a lifetime. One of your best friends even gets to go! You want to go, but your parents have said, *"No way, it's too dangerous!"* Would you rather obey your parents and stay back home until your family feels it's safe enough, or fake a letter from your parents saying you can go with your friend's family (you sneaky thing)?

LEARN TO DRAW SOME SILLY ALIEN FRIENDS!

Meet the NASA Rovers

Sojourner

Statistics

Landed on Mars: July 1997
Name of Mission: Pathfinder Mission
Length of Mission: 3 months
Size of Rover: 23 pounds (about as big as a microwave)

Mission Goals

Sojourner was sent to Mars to take pictures of its surface and send them back to NASA. It had an instrument to test rocks and soil while its lander had sensors to check the wind and weather around it. Over just 3 months, Sojourner was able to send back over 500 photographs of Mars.

Mission Details

Sojourner landed via a pyramid-shaped lander in the Ares Villis region, where scientists hypothesized that there may have been a large flood so the rover wouldn't have to travel far to examine the rocks that would have settled near its landing location. It was also a good location because it was a flat surface over which Sojourner wouldn't have difficulty traveling.

Fun Fact

This rover's name, Sojourner, is related to words like "journal" and "journey" through a root meaning "day" - the "soup du jour" is the soup of the day.

Finish the Space Monster!

Spirit

Statistics

Landed on Mars: January 2004
Name of Mission: Mars Exploration Rover Mission
Length of Mission: 3 months originally; lasted for 5 years
Size of Rover: 374 pounds (about the size of a golf cart)

Mission Goals

Spirit and Opportunity were both sent to discover evidence of water on Mars. Scientists chose the Gusev Crater as the location for Spirit because there was photographic evidence that there may have been rivers that led to the crater historically. Spirit did end up finding evidence of not only water but also volcanic, or geothermal, activity.

Mission Details
Spirit was sent with its twin rover Opportunity on a 90-day mission that was extended indefinitely, but Spirit got stuck 5 years later in 2009. Each rover landed on opposite sides of the planet so as to gain the most information. Spirit landed in the Gusev Crater.

Fun Fact
Spirit is responsible for taking and sending the first color photographs of Mars.

Draw a UFO!

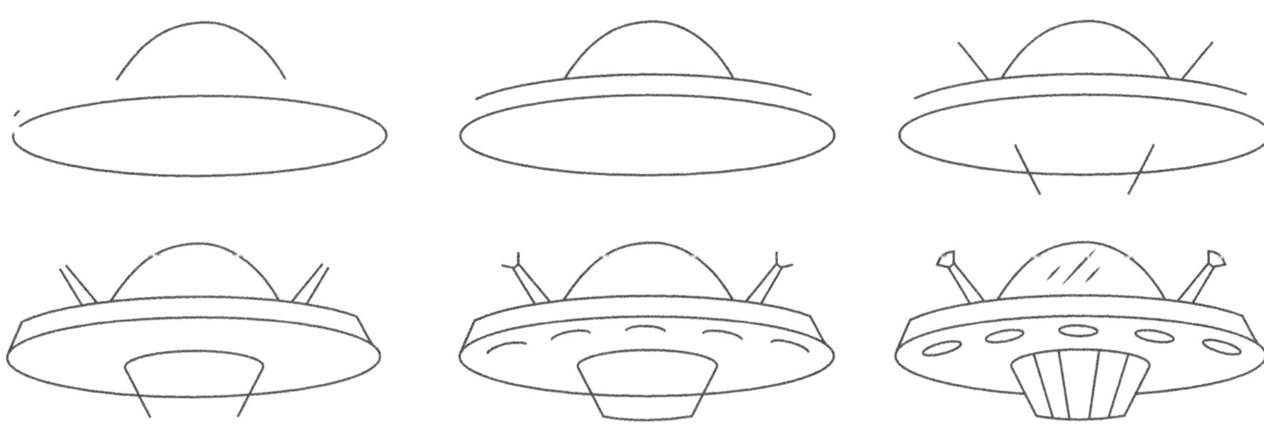

Give it a shot!

Opportunity

Statistics

Landed on Mars: January 2004
Name of Mission: Mars Exploration Rover Mission
Length of Mission: 3 months originally; lasted for 14 years
Size of Rover: 374 pounds (about the size of a golf cart)

Mission Goals

Spirit and Opportunity were both sent to discover evidence of water on Mars. Scientists sent Opportunity to the Meridiani Planum because it's a flat and wide-open area that was easy for the rover to safely traverse and because they thought there might be hematite there. On Earth, gray hematite is found near water.

Mission Details

Opportunity landed in the Meridiani Planum, on the opposite side of Mars from its twin rover Spirit. The rover quickly found what it had been sent from Earth to find: evidence of liquid water in the Martian past. After the Mars Exploration Rover mission was extended indefinitely, Opportunity survived over 5000 days by powering itself with solar energy and hibernating during dust storms to conserve that energy.

Fun Fact

Opportunity holds the rover record for longest time on Mars.

Curiosity

Statistics

Landed on Mars: August 2012
Name of Mission: Mars Science Laboratory Mission
Length of Mission: 2 years originally; extended indefinitely and is still active (as of January 2020)
Size of Rover: 1982 pounds (about the size of a small SUV)

Mission Goals

Curiosity was sent to determine whether Mars ever had the conditions needed for microbial life: a lasting water source, and certain chemicals.

Mission Details

Because Curiosity is so much bigger than the past Mars rovers, it needs wheels that are twice as big as those found on the past rovers.

Fun Fact

Curiosity has 17 cameras, one of which is attached to the end of a 7-foot-long robotic arm. The rover famously used this robotic arm as a "selfie stick" to send a photograph of itself back to Earth.

Draw a Planet!

Give it a shot!

ROCKET SHIP COLORING GRID

Use the color chart included below to color each box!
When you're done you'll have a *complete* picture you colored yourself!

20																				
19																				
18																				
17																				
16																				
15																				
14																				
13																				
12																				
11																				
10																				
9																				
8																				
7																				
6																				
5																				
4																				
3																				
2																				
1																				
	A	B	C	D	E	F	G	H	I	J	K	L	M	N	O	P	Q	R	S	T

Blue	A(1-3, 6-16), B(2, 8-16), C(8-13, 15-17), D(2-3, 10-13, 16-17), E(5-6, 11-13, 17-18), F(11-13, 17-20), G(6-7, 10-14, 16-20), H(4-20), I(4-6, 11-20), J(20), L(20), M(5-6, 11-20), N(5-20), O(6-16, 20), P(6-11, 13-15, 19-20), Q(5, 9, 10, 14-15, 19-20), R(1, 9-11, 13-16, 18-20), S(1-2, 9-20), T(3. 9-20
Black	J(7-10, 14-15, 19), K(14-15, 19-20), L(7-10, 14-15, 19)
Orange	H(1), I(1-3), J(1-6), K(1-6), L(1-6), M(1-3), N(1)

Fun NASA Facts

NASA began to operate in 1958, one year after Sputnik 1 was launched by the Soviets (modern day Russia) in 1957. Sputnik 1 was the first satellite humans had ever sent into space to orbit the Earth.

Neil Armstrong, the first person to ever walk on Earth's moon, was one week late submitting his application to NASA. His friend slipped the application into the pile so that it wouldn't look like it was late.

The Learning Channel (TLC) was founded by NASA and the Department of Health, Education, and Welfare in 1972.

NASA shows the movie *Armageddon* to new staff, then asks them to identify inaccuracies about space exploration in the film. So far, 168 have been identified.

Bill Nye the Science Guy applied repeatedly to be an astronaut but NASA has continually rejected his application.

Fun NASA Facts

There are both U.S. and Russian astronauts aboard the International Space Station. They keep separate water supplies.

The Office of Planetary Protection is a program created by NASA in case we ever discover life on another planet.

A person must travel 50 miles from the surface of the Earth before they are recognized as an astronaut by NASA.

NASA launched a space probe to Pluto in 2006. It approached Pluto 9 years later, transmitting data and photos back to Earth.

The Super Soaker squirt gun was developed by a NASA scientist.

Discover New Words: A Puzzle

Your mission, should you choose to accept it, is to discover how many new words you can create from the following phrase:

MARTIAN INVASION

1 POINT

2 POINTS

3 POINTS

5 POINTS

HOW MANY POINTS DID YOU GET?

\#______

Color this space scene!

Can you find the hidden pictures?

Ancient Discovery of the Red Planet

In ancient Rome, war and agriculture were major foundations of early Roman life and so the god of war and agriculture was an important deity in their religion, or what we would now call mythology. This deity was named Mars. Read below to discover how the planet Mars was discovered, and named after the Roman diety.

Every couple of years, Mars is closest to the Earth and thus the most visible. Because Mars can be observed with the naked eye, humans have been aware of its presence since at least the ancient Sumerians, who were around even before the ancient Greeks. Sumerians believed the Red Planet was their own god of war, Nergal, which was a trend that would continue with the planet's Greek association with Ares and its Roman association with Mars. Some of the earliest written documents describe the planet Mars as "the star of judgement of the fate of the dead."

Later, the ancient Egyptians would realize that Mars wasn't a star because it was moving through the night sky, whereas stars are consistently found in the same placement in relation to other stars -- that's why we can still find the same constellations that ancient civilizations described thousands of years ago. Planets, on the other hand, move around a star during their orbit.

Egyptians realized that Mars must be a planet, since it didn't have a stationary position like the stars in the constellations. Later in ancient Greece, Aristotle discovered that Mars was farther away than they'd thought because sometimes it disappeared behind Earth's moon, and Ptolemy made a model of Mars' orbital path. At the same time Aristotle and Ptolemy discovered such important information about Mars, the planet was also being described in ancient China by their own astronomers.

The Martian Moons: Phobos and Deimos

Like Mars, all planets in our Solar System except for Earth are named for the ancient Roman deities. The two moons of Mars, however, are named after Greek mythology.

In ancient Greek mythology, Phobos and Deimos are the twin sons of Ares (Roman: Mars) and Aphrodite (Roman: Venus). Phobos is the personification of fear, while Deimos is the personification of terror. Phobos and Deimos often accompany Ares and Eris in battle. These siblings reflect the fact that they are the children of both war (Ares) and love (Aphrodite).

Can you solve the maze?

Discover more Vocabulary Terms!

Lander: a space vehicle that is designed to land on a planet or moon

Geothermal: using the heat of a planet's interior; often volcanic

Hematite: a gray stone found near water

Telecommunications: communication at a distance, as with a telephone

Constellation: a configuration of stars into a seemingly recognizable shape (e.g., an animal or legendary hero)

Deity: one exalted or revered as supremely good or powerful, usually a god or goddess of a religion

Orbit: a path described by one body in its revolution around another, as a planet revolves around a star

Would You Rather...?

You just woke up to discover a Martian kid named Boshtix hanging out in your room, eating candy and rummaging through your clothes. After a few hours of learning to understand each other, you discover Boshtix became lost after crashing his tiny UFO, and needs to get back to Mars. He's come to seek your help. He's actually really cool, and totally hilarious! Would you rather do anything you can to help Boshtix get back home, or try to convince Boshtix to stay since he's quite possibly your future BFF?

NASA Science Division Word Search

Aerospace
Avionics
Carbon
Curiosity
Deimos
Exploration
Geothermal
Gusev Crater
Hematite
Innovation
Laboratory
Lander
Microbial
Opportunity
Pathfinder
Phobos
Propulsion
Software
Sojourner
Spirit

```
N V J T B I B L S U Y V M Z R O C
C M T P W R X V H O O B E A P F A
T P P F Z O L Y S C J S K P U E R
Q X V B S H X A I E O O O X R X B
Y D H N G V I M M B T R U A H B O
D R T X U D G U W R T I W R C A N
E C A P S O R E A U E T T P N H N
Y L F W E R E D N I F H T A P E N
Z M V Z V T S I Y O J Y T P M O R
L J X Y C W T P S H W J H O I E A
W I L N R Y S F I Y X O R S E O H
E V U M A V C Z V R B F L N Q G U
T X V D T G I U W O I U X L T I R
I G P H E L N A S D P T U N D H H
O T Y L R I O P O O I D Z F D U Z
N I R Z O I I Y R Y P F L K N U A
C R O A I R V P F L R S O M I E D
T Z T D N N A C Z M X D E C Y M G
W J A F I U G T I U D C I U W B Z
Z D R U R R B C I H M K J R B I R
L N O F M B R T E O Q P M I N L A
D C B J P O M Y H Q N G H O R L A
O H A Q B P V V Y P F M N J S N B N
M I L I L H K V G G Y F I I R B E
W C A W Y K Y V O I K L E T G H V
Y L H I N N O V A T I O N Y P Y T
L A N D E R W A N Q H R A D I H Z
X E O I W K N M I P R Z H V W B J
```

NASA Timeline Puzzle

Can you figure out what order these 8 events occurred in? Hint: they're all listed inside this book. Flip around if you have trouble remembering the answers!

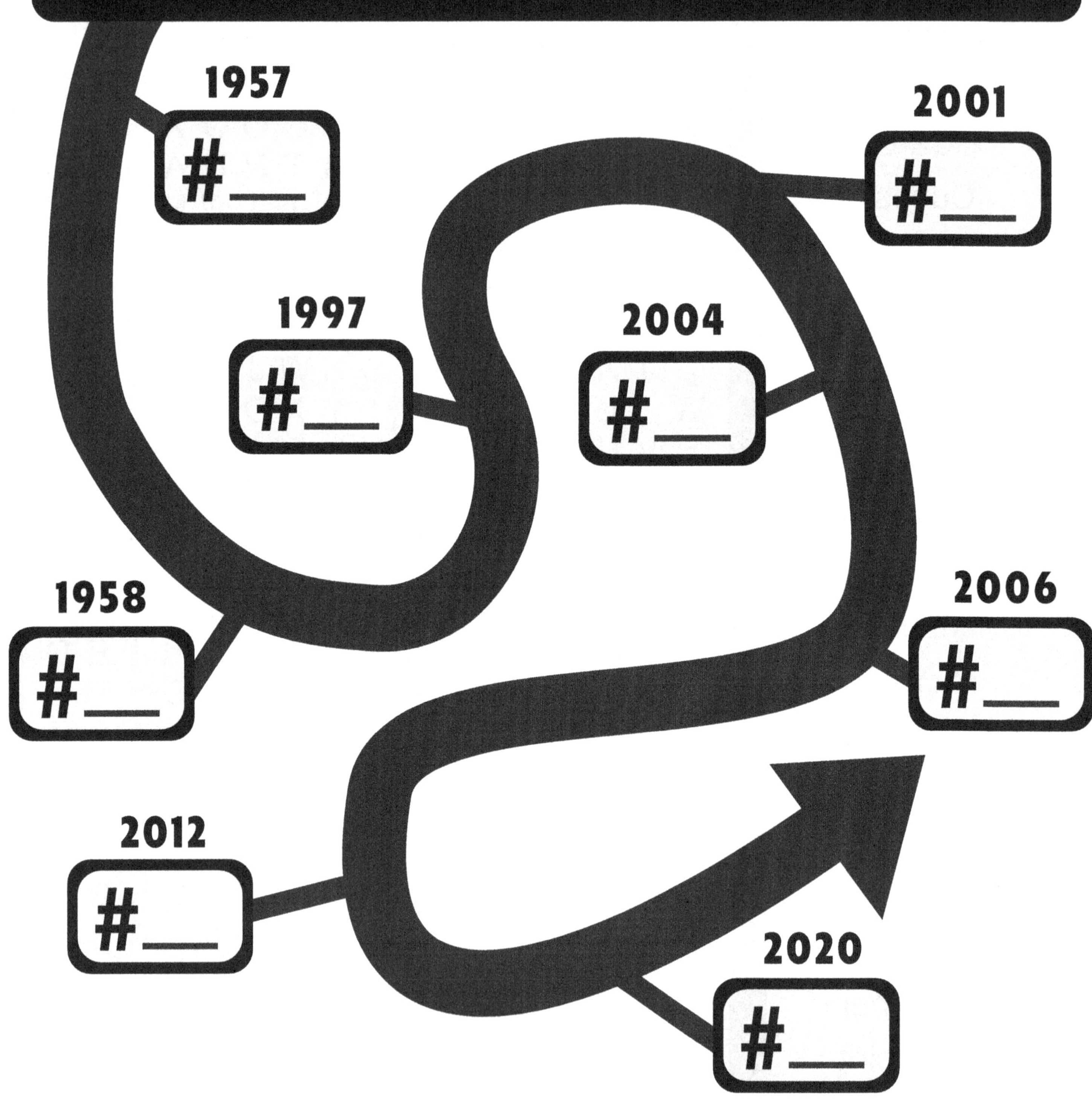

Order these events on the timeline

#1
MARS ODYSSEY MISSION

#2
MARS SCIENCE LABORATORY MISSION

#3
MARS PERSEVERANCE ROVER

#4
MARS EXPLORATION ROVER MISSION

#5
PATHFINDER MISSION

#6
NASA FIRST BEGINS TO OPERATE

#7
NASA LAUNCHES SPACE PROBE TO PLUTO

#8
SPUTNIK SENT INTO SPACE

Match the Planet

Read each trait and decide whether it describes Mars, Earth, or both. Write your answer below each trait.

Hint: all of these answers can be discovered in the pages of this book!

1. **Has 1 moon**

2. **Polar ice caps**

3. **Active volcanoes**

4. **Thin atmosphere with little oxygen**

5. **One day lasts about 24 hours**

6. **Breathable air for humans**

7. **Seasons**

8. **Changes in weather**

9. **Frequent dust storms across planet**

10. **Experiences climate change**

Answer Keys

Crosswords

page 5

Across

2. Mars 5. Discover 7. Exploration 8. Radiation

Down

1. G Force 2. Mars Rover 3. Mission 4. Evidence 6. Crater

Hidden Pictures

page 7

page 43

Mazes

page 35

page 11

page 45

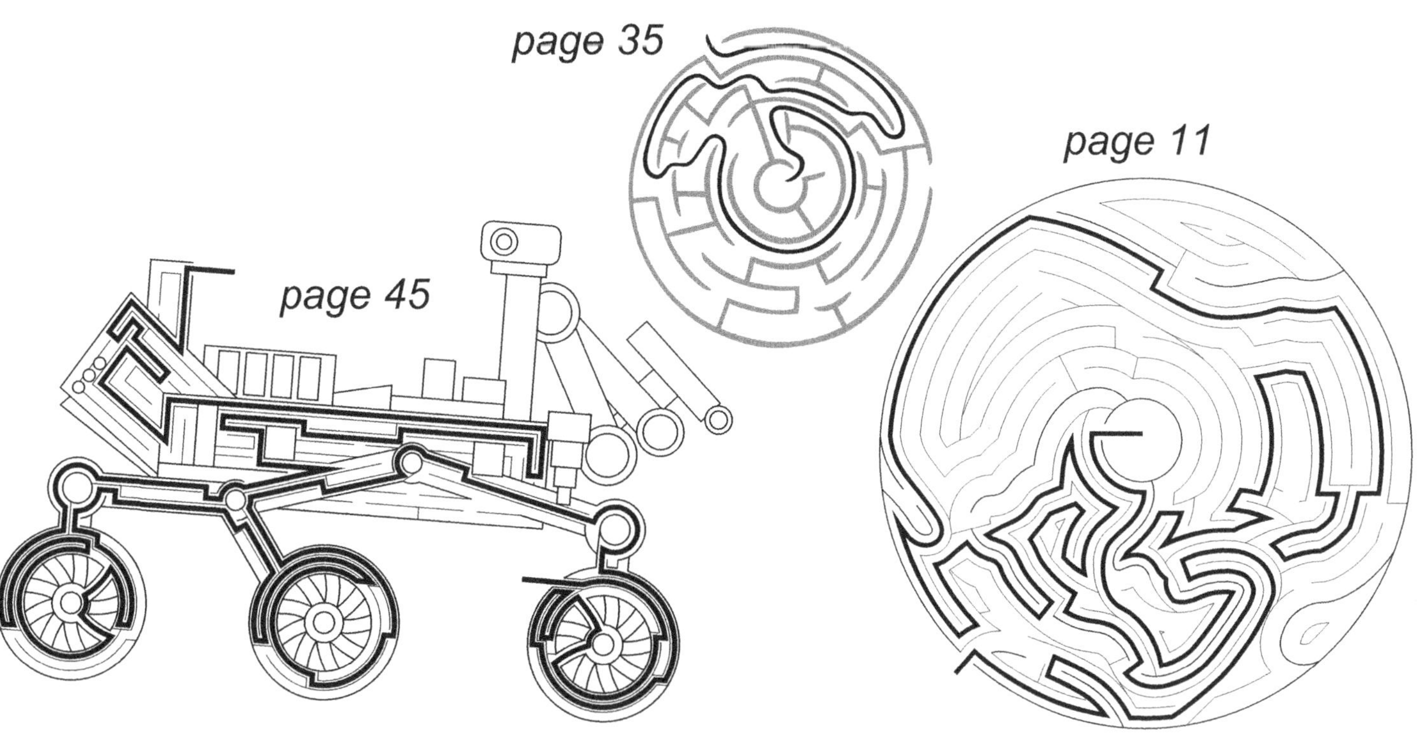

Answer Keys

Grid Coloring Pages

page 25   *page 38*

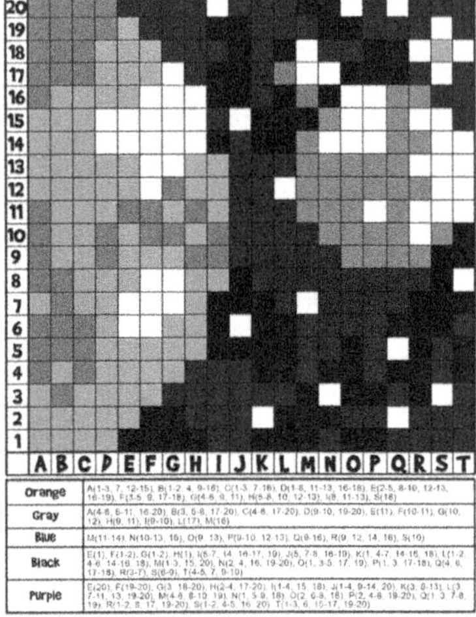

Matching Astronauts
page 16

#4 & #5

NASA Timeline Puzzle
page 48-49

| 1957: #8 | 1958: #6 | 1997: #5 | 2001: #1 |
| 2004: #4 | 2006: #7 | 2012: #2 | 2020: #3 |

Match the Planet
page 50

1. Earth   2. Both   3. Earth   4. Mars   5. Both
6. Earth   7. Both   8. Both   9. Mars   10. Both

MISSION COMPLETE

This certificate confirms that

has completed all puzzles, readings, and activities in the NASA Mars Mission for Kids, a Woo! jr. Kids space Book of Facts, Activities, and Fun!

GREAT JOB!

Made in the USA
Monee, IL
06 February 2021